Preschool Math Workbook:

Ordinal Numbers

SPEEDY
PUBLISHING

Speedy Publishing LLC
40 E. Main St. #1156
Newark, DE 19711
www.speedypublishing.com

The DUCK is placed?

1st FIRST	2nd SECOND
3rd THIRD	4th FOURTH

The **BEE** is placed?

1st FIRST	2nd SECOND
3rd THIRD	4th FOURTH

The ORANGE is placed?

1st FIRST	2nd SECOND
3rd THIRD	4th FOURTH

The **CORN** is placed?

1st FIRST	2nd SECOND
3rd THIRD	4th FOURTH

The PENCIL is placed?

1st FIRST	2nd SECOND
3rd THIRD	4th FOURTH

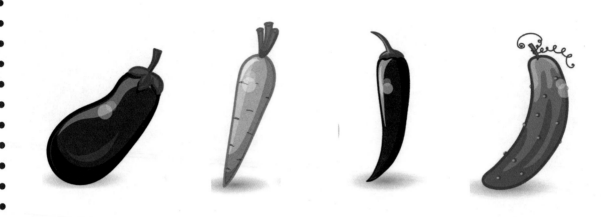

The CARROT is placed?

1st FIRST	2nd SECOND
3rd THIRD	4th FOURTH

The CORN is placed?

1st FIRST	2nd SECOND
3rd THIRD	4th FOURTH

The FISH is placed?

1st FIRST	2nd SECOND
3rd THIRD	4th FOURTH

The APPLE is placed?

1st FIRST	2nd SECOND
3rd THIRD	4th FOURTH

The **BALL** is placed?

1st FIRST	2nd SECOND
3rd THIRD	4th FOURTH

The ONION is placed?

1st FIRST	2nd SECOND
3rd THIRD	4th FOURTH

The DOG is placed?

1st FIRST	2nd SECOND
3rd THIRD	4th FOURTH

The SQUASH is placed?

1st FIRST	2nd SECOND
3rd THIRD	4th FOURTH

The EGGPLANT is placed?

1st FIRST	2nd SECOND
3rd THIRD	4th FOURTH

The **FLOWERS** are placed?

1st FIRST	2nd SECOND
3rd THIRD	4th FOURTH

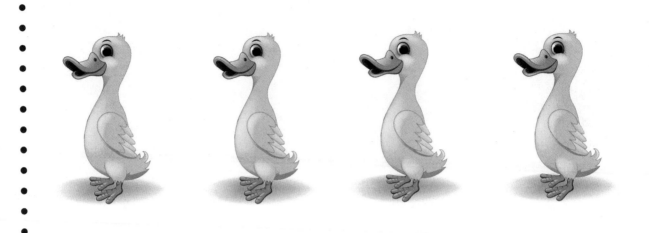

BOX the 3RD duck

CIRCLE the 1ST duck

BOX the 2ND squash

CIRCLE the 4th squash

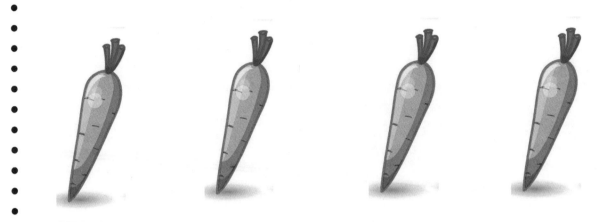

BOX the 2ND carrot

CIRCLE the 1ST carrot

BOX the 3RD ball

CIRCLE the 1ST ball

BOX the 2ND eggplant

CIRCLE the 1ST eggplant

BOX the 4TH onion

CIRCLE the 2ND onion

BOX the 3RD dog

CIRCLE the 1ST dog

BOX the 3RD strawberry

CIRCLE the 4TH strawberry

BOX the 3RD bee

CIRCLE the 2ND bee

BOX the 4TH onion

CIRCLE the 1ST onion

BOX the 2ND duck

CIRCLE the 1ST duck

BOX the 4TH squash

CIRCLE the 2ND squash

ANSWERS

3rd	2nd	2nd	1st
2nd	2nd	2nd	1st
2nd	2nd	1st	1st
4th	2nd	3rd	

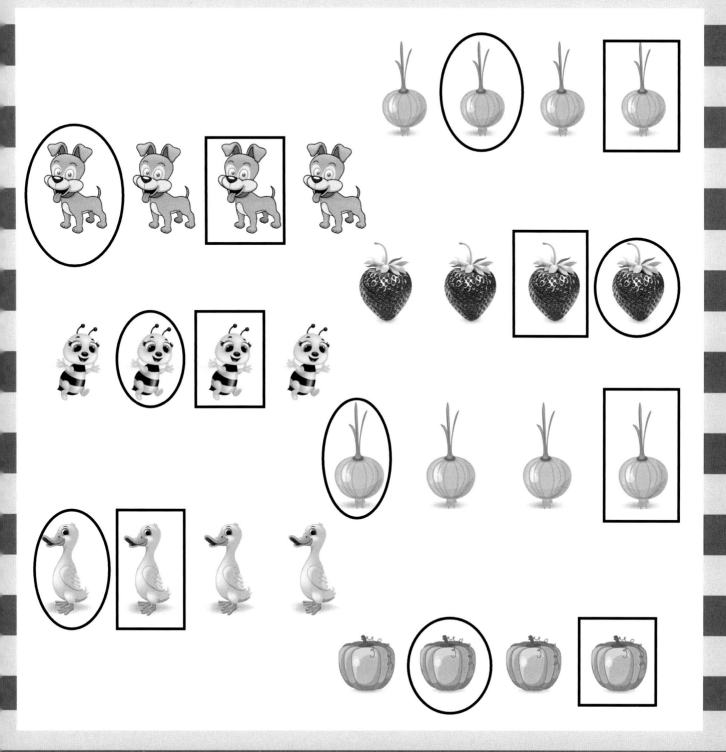

Made in the USA
Monee, IL
14 October 2023

44615280R00021